2024

TURBO TABS
IFGC®

- Tabs for the Code sections you reference the most
- Durable, self-adhesive tabs
- Use with soft-cover edition of the Code or Commentary

Turbocharge your code!

ISBN 978-1-962103-07-7

Item No. 0601TS24

ICC
INTERNATIONAL CODE COUNCIL®

Installation Instructions for *Turbo Tabs*

1. Please read instructions carefully **before** applying adhesive tabs as they will stick **permanently** on first contact. All tabs must be placed on the **right-hand page**.

2. Begin first tab application by placing the first printed, **BOTTOM RIGHT** tab in the **back** of the code. Align vertical line on tab with the edge of the page. Position this tab approximately **½ inch** from the **bottom** of the page, then apply, as shown in **Example 1** below.

3. Apply the next bottom right, consecutive tab on the respective page. Overlap this tab slightly with the previously applied tab, as shown in **Example 2** below. Continue applying the consecutive remaining tabs, working **up** toward the top of the code. If you reach the top of the book, start again at the bottom with another row.

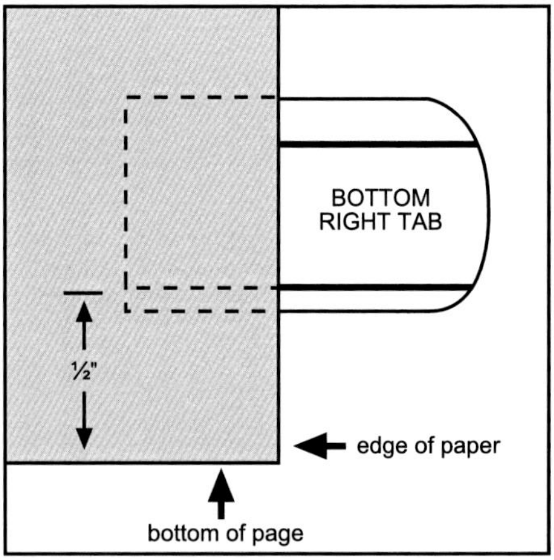

Example 1

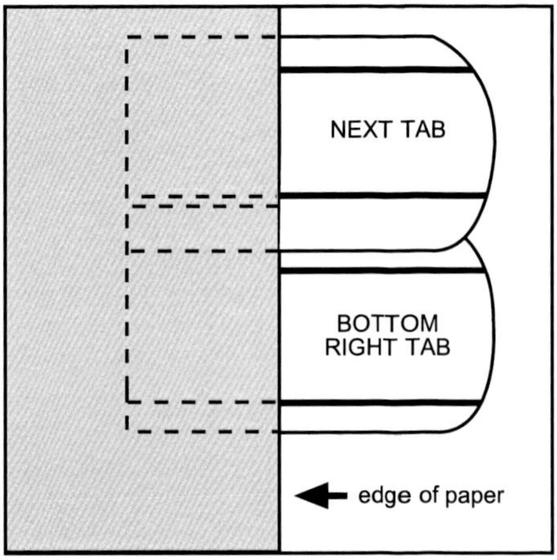

Example 2

2024 INTERNATIONAL FUEL GAS CODE® TABS INSERTION INSTRUCTIONS

COMPLETE ITEM DESCRIPTION	ABBREVIATED TAB DESCRIPTION	CHAPTERS SECTIONS TABLES
Scope and Administration	SCOPE & ADMIN	Chapter 1
Inspections and Testing	INSPECT & TESTS	Section 111
Definitions	DEFINITIONS	Chapter 2
General Regulations	GENERAL REGS	Chapter 3
Combustion, Ventilation and Dilution Air	COMBUST AIR	Section 304
Access and Service Space	ACCESS	Section 306
Reduction of Clearances with Specified Forms of Protection	CLEAR REDUCT	Table 308.2
Gas Piping Installations	GAS PIPING	Chapter 4
Piping Materials	PIPING MATLS	Section 403
Piping System Installation	PIPING SYS INSTALL	Section 404
Inspection, Testing and Purging	TESTING & PURGING	Section 406
Shutoff Valves	SHUTOFF VALVES	Section 409
Support of Piping	PIPING SUPPORT	Table 415.1
Chimneys and Vents	CHIMNEYS & VENTS	Chapter 5
Venting of Appliances	VENTING	Section 503
Type of Venting System to Be Used	VENT TYPES	Table 503.4
Gas Vent Terminations	GAS VENT TERM	Section 503.6.5
Through-the-Wall Vent Terminal Clearance	VENT TERM CLEARANCE	Table 503.8
Clearances for Connectors	CONNECTOR CLEARANCE	Table 503.10.5
Maximum Vent Connector Length	VENT CONNECTOR LENGTH	Table 504.3.2
Factory-Built Chimneys	FACTORY-BUILT CHIMNEYS	Section 506
Specific Appliances	APPLIANCES	Chapter 6
Clothes Dryer Exhaust	CLOTHES DRYER EXHAUST	Section 614
Forced-Air Warm-Air Furnaces	FORCED-AIR FURNACES	Section 618
Cooking Appliances	COOKING APPLIANCES	Section 623
Gaseous Hydrogen Systems	GAS HYDROGEN SYSTEMS	Chapter 7
Testing of Hydrogen Piping Systems	HYDROGEN SYS TESTS	Section 705
Referenced Standards	REFERENCED STDS	Chapter 8
Sizing and Capacities of Gas Piping	GAS PIPE SIZING	Appendix A
Sizing of Venting Systems Serving Appliances Equipped with Draft Hoods, Category I Appliances and Appliances Listed for Use with Type B Vents	VENT SYS SIZING	Appendix B
Recommended Procedure for Safety Inspection of an Existing Appliance Installation	SAFETY INSPECT EXIST APPL	Appendix C
Board of Appeals	BOARD OF APPEALS	Appendix D

CHAPTER 1 — SCOPE & ADMIN

SECTION 111 — INSPECT & TESTS

CHAPTER 2 — DEFINITIONS

CHAPTER 3 — GENERAL REGS

SECTION 304 — COMBUST AIR

SECTION 306 — ACCESS

TABLE 308.2 — CLEAR REDUCT

CHAPTER 4 — GAS PIPING

SECTION 403 — PIPING MATLS

SECTION 404 — PIPING SYS INSTALL

SECTION 406 — TESTING & PURGING

SECTION 409 — SHUTOFF VALVES

TABLE 415.1 — PIPING SUPPORT

CHAPTER 5 — CHIMNEYS & VENTS

SECTION 503 — VENTING

TABLE 503.4 — VENT TYPES

SECTION 503.6.5 — GAS VENT TERM

TABLE 503.8 — VENT TERM CLEARANCE

TABLE 503.10.5 — CONNECTOR CLEARANCE

TABLE 504.3.2 — VENT CONNECTOR LENGTH

SECTION 506 — FACTORY-BUILT CHIMNEYS

CHAPTER 6 — APPLIANCES

SECTION 614 — CLOTHES DRYER EXHAUST

SECTION 618 — FORCED-AIR FURNACES

SECTION 623 — COOKING APPLIANCES

CHAPTER 7 — GAS HYDROGEN SYSTEMS

SECTION 705 — HYDROGEN SYS TESTS

CHAPTER 8 — REFERENCED STDS

APPENDIX A — GAS PIPE SIZING

APPENDIX B — VENT SYS SIZING

APPENDIX C — SAFETY INSPECT EXIST APPL

APPENDIX D — BOARD OF APPEALS

2024 INTERNATIONAL FUEL GAS CODE® TABS